Ferns for Modern Living

Contents

Publisher's representative for distribution in Canada. Horta Craft Ltd., London, Ontario

ISBN 0-89484-004-5

Acknowledgments

Elaine Davenport has been a fancier of ferns and other houseplants for years. Her collection numbers as many as 300 different plants. Elaine has her degree from Western Michigan University and has done graduate work there as well. She teaches Care of Indoor Plants at St. Clair County Community College in Port Huron, Michigan. Elaine holds membership in the Los Angeles International Fern Society, The American Fern Society and the American Horticultural Society. She and her husband, two daughters and cat live in Imlay City, Michigan.

John Pike started his photographic career at the age of fifteen with a darkroom set up in his parents' basement. He has handled almost every area of photography from Chief Photographer for Civil Defense to photographing for *Motor Guide* magazine. John's photographs have appeared in numerous publications and catalogs as well as in Merchants' MODERN LIVING book series. Recognized for the high quality of his photography, John has been honored several times by professional photographers' associations. In the past few years he has travelled extensively for Merchants, photographing plants all over the United States. John, his wife and family reside in Kalamazoo, Michigan.

The text for *Ferns for Modern Living* was written by Elaine Davenport. Guidance and assistance is acknowledged with thanks to David Roberts of Select Nurseries, Inc., and M. Jane Coleman Helmer, Ph.D. of Merchants Publishing. Edited by Helen VanPelt Wilson.

The photographs are from Merchants' comprehensive library of horticultural subjects; a collection of more than 15,000 pictures compiled by photographers John Pike and the late Willard Kalina.

We thank the following companies and individuals for their kind cooperation in providing assistance, materials, and plants for use in this book:

Belle Isle Greenhouse, Belle Isle Park, Detroit, Michigan; Fernwood, Inc., Niles, Michigan; The Green Thumb, Kalamazoo, Michigan; Green Thumb Products, Apopka, Florida; Riverside Greenhouses, Kalamazoo, Michigan; Tri Tiki Nurseries, Winter Garden, Florida.

Hardy Ferns to enjoy outdoors

Ferns offer a wide variety of plants for different uses. They vary greatly in size, pattern, texture and shades of green. In planning your outdoor garden, consider the type of ferns to be used and the space you will want them to occupy. Many spread rapidly while others grow slowly. Some stand like sentinels, rising to more than 4 feet, while others are low growing. The Maidenhair is dainty and fan-shaped. The Christmas fern is lance-shaped and leathery. Many, such as the Autumn fern, have a reddish color when the fronds are young. The Japanese Painted fern is perhaps the most chromatically unique of all ferns with its wine-red and soft silver-grey fronds.

Ferns can soften a landscape. Group them by size, set large plants along a fence, porch, or foundation line, miniature varieties in rock gardens, stone walls, along walkways and borders. Plant spreading varieties as groundcover for grass-less areas under trees or shrubs. Use ferns anywhere to enhance the beauty of flowering plants or where attempts to grow flowers in a sunless north exposure have failed.

Ferns are a natural for the country home. A brook or pond almost begs for a planting of ferns along its banks, or you might create a foot trail. An excellent way to map out a trail is to walk a winding route, allowing a heavy string to unwind behind you. Take advantage of large stones, boulders, and fallen logs for your fern-planting sites. Select a spot or two for a resting place. Mark it off with fallen logs or railroad ties and place a rustic bench. In this area, plant several kinds of ferns and mix them with wild flowers — wood violet, columbine, trillium, blood-root, May apple. Often an interesting fern will be found tucked away in an inacces-sible area, so transplant it into your view. Your trail need not be extensive; its appeal will derive from the harmony with nature that you have created.

Transplanting ferns is a simple procedure. If you are acquiring ferns by this method and the property is not your own, be sure and obtain the owner's permission. Take care to dig ample soil from around the roots, and cover the uprooted plants with wet toweling for transport. Don't leave them in a hot closed car. Spring is a good time for transplanting.

Since many ferns naturally exist in cool, moist, wooded areas, try to approximate these conditions if a specimen garden is desired. Most ferns need soil that is light in texture. Remove poor soil, especially clay, to a depth of 8 to 10" (20-25cm) and replace it with a mixture containing equal quantities of topsoil, humus, peatmoss, leafmold, and sand. Soil prepared in this way will aid water retention and aeration. Addition of extra leafmold, principally oak leaves, will increase acidity. Most ferns prefer a soil that is slightly acid, a pH of 6. For those liking an alkaline soil, add commercial lime, crushed limestone rock, or oyster shells. To conserve moisture, mulch the fern bed year round. Mulching can be done naturally and inexpensively with leaves, shredded bark, or larger pieces of bark from a fallen, partially decomposed tree.

The maintenance of ferns is no more difficult than for other plants you prize. They are almost insect resistant. Sometimes snails or slugs can be a problem on low-growing varieties; if the problem persists, treat with a commercial slug bait.

Ferns rarely look their best planted in direct sun, and a cool shaded place makes watering less frequent. In sunnier locations, water more frequently as you would a mixed bed of flowers. For best results in wintering the hardy growers, mulch in the fall with a generous amount of dry leaves or straw. Await the appearance of the croziers, the young fronds, in spring, a real delight, before removing most of the mulch.

Most of the hardy ferns pictured in this book can withstand below-freezing periods and are, therefore, suitable for areas where winters are long and severe as in the Northeast, Northwest, Southeast and Central regions of the continental United States. The Southwest presents a unique problem because of heat and aridity. However, these hardy species are adaptable even to this area:

Lady fern, *Athyrium filix-femina;* English Male fern, *Dryopteris filix-mas;* Sensitive fern, *Onoclea sensibilis;* Christmas fern, *Polystichum acrostichoides.*

In southern Florida and Hawaii a wide range of ferns usually considered as houseplants or tender species can be grown. The main requirements for fern growing there are mild winters and humid summers. The Bird's-Nest fern, the Squirrel's-Foot fern and many varieties of the Boston fern can be successfully added to the outdoor landscape. Likewise, for the temperate, coastal climates and for parts of southern California, but not the inland, desert and mountain areas, a variety of tender ferns can become outdoor specimens. Holly ferns and tree ferns do well. A host of possibilities are open for ferns in milder, temperate regions. More than 100 species are hardy outdoors along the coast from just north of San Francisco to San Diego. Only a few have been named for you here. Generally, the ferns grown in these climates withstand considerable sunlight, especially the Sword fern, Trembling Brake, and Australian tree fern.

Indoor Ferns to start with

When you mention ferns to almost anyone, the Boston fern immediately comes to mind. Of all the potted ferns, it is the most popular. However, as you will discover in this book, it is not the only beautiful fern for your home.

Don't equate popularity with ease of growing. While it is still a good choice for the beginner, the Boston fern is not necessarily the easiest. Problems arise if it is not given sufficient root moisture, good light with some winter sun, and adequate humidity (at least 40-50%).

Listed below are a few ferns beside the Boston that are of relatively easy culture. You may wish to start with one of these, or add it to your plant collection. You will find no Maidenhair ferns listed here, although the Australian maidenhair fern *(Adiantum hispidulum)* is one of the easier varieties. Generally, Maidenhair ferns require more humidity (above 50%) than the average home can supply. Ferns listed here are "easy" because they have one or more of these characteristics:

- tolerate lower humidity (about 30% or less)
- tolerate lower light intensities
- tolerate winter sun
- tolerate occasional drying out without collapse, especially during winter

1. Holly — *Cyrtomium genus*
2. Button — *Pellaea rotundifolia*
3. Bird's-Nest — *Asplenium nidus*
4. Staghorn — *Platycerium genus*
5. Rabbit's-Foot — *Polypodium genus*

As you look through the pages of this book and discover a fern that appeals to you, check its growing conditions to see if they can be matched in your home. Often if you can provide higher humidity, you can grow that so-called "difficult fern".

Basic Culture

Ferns have the same basic requirements as other plants; they do not need a Victorian parlor or a tropical greenhouse in which to thrive. A green thumb has little to do with fern culture, but understanding the plants' needs and fulfilling them does. That understanding can be broken down into seven requirements — light, temperature, humidity, water, soil, potting, and fertilization. It is important to remember that, although ferns are prized for their decorative value, they are not objects, but living organisms.

Light-Where Will My Fern Thrive?

Plants require light. A common misconception is that ferns will grow with little or no light, as in a dim room. The amount of light required depends largely on the type of fern. Generally, ferns grow best in filtered or diffused light, that is, in a bright position out of the direct rays of the sun, or at or near a window where the rays are diffused and filtered by a sheer curtain. While ferns will not tolerate strong summer sun, some will accept an east or west window in summer. In fact, in the North in winter, a full southern exposure is beneficial for some ferns since the sunlight is usually of shorter duration and weaker intensity. Northern windows, except perhaps a bright, unobstructed one in summer, are not bright enough for most ferns. Ferns receiving too much light are less luxuriant, lose their color, and may develop brown leaf margins. Ferns receiving too little light may fail to grow or may produce tall, spindly, undersized fronds.

Light Requirements

Direct Light or Sunlight (high to medium-high light) sun falling on all or part of the leaves. This usually occurs for a few hours a day depending on window exposure. The majority of ferns prefer some protection from full midday and afternoon sun during the summer.

Filtered, Indirect, Diffused (medium-high to medium-low light) a well-lit position out of direct rays, or a position with a sheer curtain between fern and sun. This type of light can occur at any window, depending on obstruction. Most ferns grow well in this light situation.

Semishade or Shade (medium-low to low light) a postion at a distance from a window, perhaps on a table or stand. Here ferns have to rely heavily on reflected light, so a room with light-colored walls is best.

Light According to Window Exposure

North — least light and heat of the four exposures. Good for very low-light ferns.
South — longest and most intense source of direct light and heat. Not suitable for

most ferns, unless light is filtered by a curtain and humidity and ventilation are adequate. Good source of winter light for ferns in colder, northern climates.

East — direct light that is cooler than from south or west, and therefore less dehydrating. Good year-round exposure for ferns.

West — direct light that is about the same duration (4 hours) as eastern exposure, but hotter. Protect ferns in summer months by filtering with a curtain. From November to March, in colder climates, it is not necessary to use curtaining to reduce the heat and light intensity from this exposure.

Ways to Modify Light

Light not only varies by window exposure, but the quality of light also varies from room to room. Light-hued walls reflect more light than walls painted in darker tones, white being the most effective. Wallpaper with a foil-like quality can reflect light in almost the same way as a mirror. To improve the quality of the natural light for your ferns:

- open curtains
- prune shrubs and trees close to windows
- keep windows and plants free of grime and dust
- consider effect of awnings or overhangs as you place your ferns
- hang sheer curtains at appropriate windows and in appropriate seasons
- consider the coverings and color of walls
- hang mirrors
- allow sufficient spacing of plants

Temperature

Ferns vary considerably in their temperature requirements, and these usually depend on the habitat, that is, the native environment of the species. Ferns grow not only in the tropics but also near the Arctic and Antarctic. Important as it may be to establish relative guidelines to temperatures, the fern as a houseplant no longer exists as a "natural creature" of woodlands and tropics. It has been grown for you in a greenhouse. As a houseplant, it will grow best for you in a range of 65°F to 80°F (19-27°C) with a 10°F (5°C) drop in temperature at night. Requirements can be further refined to cool, average, and warm household climates.

Cool preference ferns do best with day temperatures of 60°F to 70°F (16-21°C) while night temperatures may go as low as 45°F (7°C). These ferns are sometimes grown out-of-doors in milder climates and can be considered semihardy. Ferns capable of tolerating temperatures below freezing can be considered hardy.

Average temperature means a daytime range of 65°F to 70°F (19-21°C), even up to 75°F (24°C). The usual night requirement is 50°F (10°C). Ferns with this tolerance are considered semitender; if kept out-of-doors below the minimum night temperatures, their foliage may be severely damaged.

Average-to-warm temperature ferns need daytime heat of 70°F to 75°F (21-24°C) even up to 80°F (27°C). Nights should not go below 60°F (16°C) or growth will be poor. These ferns, generally known as tender or tropical, flourish year-round in warmth and humidity.

While ferns are reasonably adaptable and tolerant of slight variations in temperature, a uniform day-night cycle is best. Excessive heat or drafts, especially around windows in winter, are fatal to ferns as they are to most foliage plants. So it is advisable to consider all this in the placement of ferns in your home. Remember that air temperature is warmer close to the ceiling and generally cooler near the floor. On this account hanging plants must be watered more frequently than table or floor plants since the heat and the air movement tend to dry out the soil.

Humidity

Humidity is important to the health of your ferns. As a rule, ferns do not thrive in low humidity, but some require less humidity than others. Generally, a range of 30% to 60% is necessary.

To increase humidity:

- mist twice a day with tepid water
- group plants
- set pots on pebble- or gravel-filled trays with water. Be sure pots rest on the pebbles or gravel above the water line, not in the water
- place in kitchen or bathroom where humidity is generally higher than in other rooms
- supply a humidifier

Ferns requiring less humidity than others include: Holly ferns, Bird's-Nest, Rabbit's-Foot or Hare's-Foot, some Staghorns, and table ferns, like the Trembling Brake.

Most Maidenhair ferns, but not the Australian Maidenhair, the Boston ferns, and Tree ferns require high humidity.

Watering

No schedule, such as once, twice, or three times a week should be applied to watering your ferns. Watering depends on a combination of factors — temperature, light, humidity, density of soil, type and size of pot, as well as the kind of fern and the time of year.

Most ferns prefer to be just moist, but not wet. Others require less moisture, and some will tolerate an occasional drying out without collapse of foliage. Properly moist soil should look and feel damp. If soil is dry ½ to 1" (1.8-2.5cm) down from the top, it is time to water again.

How Much Water Should I Give My Ferns?

Whenever you water, be thorough so that the **whole** root ball (all the soil) is moistened and excess water seeps out the drainage hole. If the pot has no drainage holes, lay it on its side, if possible, and allow excess water to run off. Ideally, every pot saucer should be checked about half an hour after watering, and any surplus water emptied out.

Most ferns dislike wet feet. Roots need oxygen for life and growth. If plants are overwatered the oxygen is pressed out as the saturated soil mass packs down. Eventually the roots die, usually from rot. If the soil is still wet three or four days after watering, feels soggy, and smells sour, or if the pot is unusually heavy for its size, you may be overwatering your fern.

Is Tap Water O.K.?

Ordinary tap water can be used, but cold water shocks plants. Don't use hot water either. Tepid water — water that feels comfortable to the hand — is best. Add a little warm water to the cold, or let water stand overnight so that it reaches room temperature.

Chlorine in tap water does not usually harm ferns. Chlorine is an unstable element. Allowing water to stand in an open, wide-mouthed container for 12 to 24 hours will permit most of the chlorine gas to escape.

Methods

From the top—

The simplest way to water a plant is from the top, letting water drain out the bottom. An advantage to this method is the purifying of the soil (leaching) through removal of excess salts and fertilizers. If your fern is in a clay pot, you probably have evidence of this purifying process. White crusts around the outside of the pot are the result of excess salts and fertilizers. If you are using a plastic or decorative container, it may be necessary to leach out the excess salts and fertilizers periodically by immersing the potted plant in tepid water for 30 minutes, or until bubbles stop rising. White crusts on the surface of the soil are an indication that your fern needs to be leached of harmful salts.

From the bottom—

Large ferns, difficult to water from the top, may be placed in water **up** to the rim of the pot, and left long enough for the soil to be thoroughly moistened, usually a period of 15 to 20 minutes.

Alternatively, you may set the clay pot in a saucer, (clay or otherwise), add water to this saucer, and continue to add water until the soil is moistened by capillary action and feels damp on top. Ferns that are bottom watered need to be leached occasionally.

Treating A Wilted Fern

If your fern has wilted from lack of water, water it immediately and keep watering even after the first drainage of water. You want to make sure that the water is penetrating the root ball, not merely running between the sides of the pot and the dried and slightly shrunken soil mass. To guarantee a thorough watering, submerge the potted plant in water for 15 to 20 minutes. Then place the fern in subdued light, mist the foliage several times through the day, and wait for the plant to revive. It should do so within 24 hours. Later, remove any fronds that do not survive.

Treating An Overwatered Fern

Yellowing of foliage, other than as a result of natural aging or disease, is usually a sign of overwatering and/or poor drainage. Re-evaluate your watering regimen and check soil and pot for aeration and drainage.

If the fern has wilted from overwatering, it is necessary to knock it out of the pot to check the health of the roots. If these are soft and discolored, the plant may not survive. You can try cutting off the damaged roots, repotting in fresh soil that is well aerated and usually moving to a smaller pot. The plant will probably look miserable; give it a chance to rejuvenate, but don't expect a miracle. Try not to overwater your ferns again!

Fertilizing

Ferns generally need little fertilization to maintain normal growth because the humus-rich soil provides the necessary nutrients. However, you may want to replenish the soil with nutrients. This can be done with a single application of fertilizer every 4 to 6 months. Or you may want to match fertilization with the active growing period — usually spring and summer — by giving applications of fertilizer every 2 to 3 weeks. Resting or inactive plants do not need fertilization. Guidelines for fertilizing are given under Culture for each plant.

Choosing a Fertilizer

Any complete fertilizer containing the elements nitrogen, phosphorus, and potassium (N-P-K) is fine, but no fertilizer is good if applied too strong. Strong fertilizers can burn the fragile roots of ferns. Try to select one with a reputation for "low burn;" fish emulsion, 5-2-2, is an excellent choice. Dilute any other complete fertilizer to ¼ the recommended strength before applying it. Never apply to dry soil. Water the plant thoroughly; then fertilize.

Bone meal, 0-11-0, is an incomplete fertilizer added to the layer of soil below the roots to promote root growth. It also has low-burn qualitites. It is not necessary to replenish it until the plant is repotted. However, a complete fertilizer should be used in addition to the bone meal.

Liquid, or water-soluble fertilizers, are easy to apply and there is little chance of burn or over-fertilization with them if they are used properly. Dry fertilizers in the form of granules that are spread over the surface of the soil are not recommended. If a dry fertilizer contains much soluble salt, and this is accidently concentrated in one spot, the result may be fertilizer burn and damage to the plant tissues.

Soils

To perform best, ferns need a soil suited to their particular needs. Information that accompanies each illustration in this book includes a note on the preferred soil.

Generally speaking, packaged soils, although a good base for your fern mix, do not contain sufficient humus. Nor do these mixes usually contain enough sand or perlite for adequate drainage. To insure healthy plants and lush growth, add some or all of the following materials to your soil mix:

Additives

Humus, derived from decayed vegetation, provides aeration and essential acidity. It also increases the water-holding capacity of the soil. Peatmoss and leafmold are familiar forms.

1. Peatmoss refers to sphagnum peatmoss. It may be partially or wholly decomposed. Unmilled or natural sphagnum strands are useful to line wire baskets. Chopped or milled sphagnum moss is incorporated in the soil mix. The less-decayed, coarse, light-brown peat is good for ferns. Do not use the wholly decayed, velvety black peat that is sold for dressing on lawns. Dried sphagnum moss and peatmoss are initially difficult to moisten and both will repel water if not pretreated. To pretreat, pour boiling water over the peat, let it soak; wring out the excess water after it has been allowed to cool. Then the peat is ready to be mixed with the soil.
2. Leafmold, partially decayed leaves, is usually sold as oak leafmold, It is the most popular humus for ferns.

Charcoal absorbs toxic gasses and salts caused by poor aeration in water-logged soils. Be sure to purchase a horticultural grade of charcoal. Place a layer at the bottom of the pot; also add it to the general soil mix.

Perlite, white, sterile, volcanic rock, lightens the soil, improves drainage and aeration. Use medium or coarse grades. Sharp sand can be substituted for perlite.

Sand improves drainage and aeration of soil. Use sharp sand or builders' sand, if this is not too fine, or washed concrete sand.

Mix **oyster shells**, dolomite limestone, or horticultural lime in soil for ferns liking a basic (alkaline) soil. Oyster shells also provide extra aeration.

Bonemeal is a phosphorus-rich fertilizer to add to the soil at the time of potting. It promotes root growth.

Basic soil mix:

1 part packaged general houseplant soil; 1 part humus; 1 part sharp sand or perlite. Add ½ teaspoon of bonemeal to the soil mix under the roots when potting or repotting.

Humus-rich mix:

1 part packaged general houseplant soil; 2 parts peatmoss; 1 part leafmold or ½ part ground redwood chips and ½ part oak leafmold; 1 part sharp sand or perlite. (To decide whether your particular ferns need acidic or basic soil, refer to culture for each one. Add a light dusting of dolomitic limestone or about 5 tablespoons of crushed oyster shells for those plants requiring a basic soil.)

All soils need a combination of water-holding and water-draining material. These two formulations should give you that balance.

Asparagus Ferns

Ask almost anyone and you will be told it's a fern. Actually with its feathery foliage the asparagus fern is a nonedible relative of the garden asparagus, and a member of the lily family. *Asparagus densiflorus 'Sprengeri'* bears seeds, botanically differentiating it from the true ferns that reproduce by spores, and has small six-petaled pinkish white flowers that develop into red berries the size of peas. Yet, the foliage is as delicate looking as that of a maidenhair fern.

Asparagus ferns are fast-growing, have massive adventuresome roots, and are greedy, so the plant you buy from the florist is probably already potbound or soon will be. Be prepared to give it a new home, and fertilize it every two weeks; new shoots will appear almost immediately. Generally, the plant likes to be kept moist and cool, but it can be allowed to dry out and watered thoroughly again with no great harm. Plants adapt to strong winter sunshine and to strong light indoors in all seasons. However, strong sun plus excessive heat may cause the plant to turn yellow and lose its refreshing cool green color.

PLUMOSA FERN
Asparagus setaceus

Flat, dark green, fernlike growth in a horizontal plane on a stem that tends to climb. Because of durability, often used by florists in bouquets and table decorations. Grows in sun or shade.

ASPARAGUS FERN ▸

Asparagus densiflorus 'Sprengeri'

Fern look-alike for hanging baskets is this lacy trailer with a spray of fine needlelike leaves. Small, white blossoms ripen into coral-red berries when plant is mature. Good for an area too dry for regular ferns. Bulbs in root system store water, making this "fern" comparatively drought-resistant.

Culture: Asparagus Ferns

Temperature: Average to cool; minimum night 45-55°F (7-13°C).

Light: Bright or filtered light; some sun, especially during winter enhances growth. Too much sun or too hot a location will cause needles to yellow and drop.

Watering: Prefers moist soil, but will tolerate short period of dryness.

Fertilizing: Every 2 weeks year-round with ¼ the recommended strength.

Soil: General houseplant mix. Provide good drainage.

Special Consideration: Asparagus and plumosa ferns may outgrow containers. To keep within limits, it may be advisable to root-prune. Use a sharp knife and cut away some roots from bottom and sides before repotting.

MING FERN ▸

Asparagus densiflorus

This fern mimic with Far Eastern flare offers small puffs of bright green needles; 8-10′ (2.4-3m) specimens can be seen outdoors in Southern California.

FOXTAIL FERN ▸

Asparagus densiflorus 'Myers'

Highly decorative upright grower with dense, fluffy plumes. Expect slow, but steady growth.

Maidenhair Ferns

Adiantum 'Birkenheadii'
Similar to the Australian but with fronds erect and lighter green. Dependable for home cultivation.

Culture: Maidenhair Ferns

Temperature: Average climate, minimum at night 55°F (13°C); will tolerate temperatures of 45°F (7°C) for short periods. Optimum daytime temperature is 70°F (21°C).

Light: Low to medium; protect from direct sun. Good in unobstructed north light or in filtered light of an east window. Prefers humidity range of 40% to 60%. Curling and browning of new fronds may be an indication of too low humidity.

Watering: Keep soil uniformly moist but not wet. Avoid overwatering. In winter, allow soil to become slightly dry before rewatering, but do not allow plant to wilt.

Fertilizing: Only during active growth with dilute solution every 3 weeks or space applications every 4-6 months according to growth.

Soil: Add peatmoss, leafmold, or other humus to a houseplant mix. Sharp sand or perlite will add drainage. Most Maidenhair ferns like calcium-rich soils, so add dolomite limestone or oyster shells to the mix.

Special Considerations: Most Maidenhairs have a dormant period in winter, and it is advisable to rest them during this time. Cut off all old and discolored fronds to make way for new growth in spring. Reduce watering and set plant in a cool place. In spring, repot or top dress with fresh soil mixture. When small, these ferns are best used in terrariums.

Caution: Maidenhair ferns can be damaged by some insect sprays. See Pests and Disease, page 75.

Adiantum tenerum 'Scutum'

Fronds are tinged red with new growth more vivid than older leaflets.

Adiantum raddianum 'Fritz-Luthii'

Spirals of steel-blue shingled leaflets are characteristic of this vigorous grower.

AUSTRALIAN MAIDENHAIR *Adiantum hispidulum*

Easy to grow, the five-fingered fronds are rosy-bronze when young, changing to medium-green when mature.

MAGNIFICIENT MAIDENHAIR
Adiantum tenerum farleyense

Deep rose contrasted with pea-green make the fluted leaflets of this plant a chromatic pleasure.

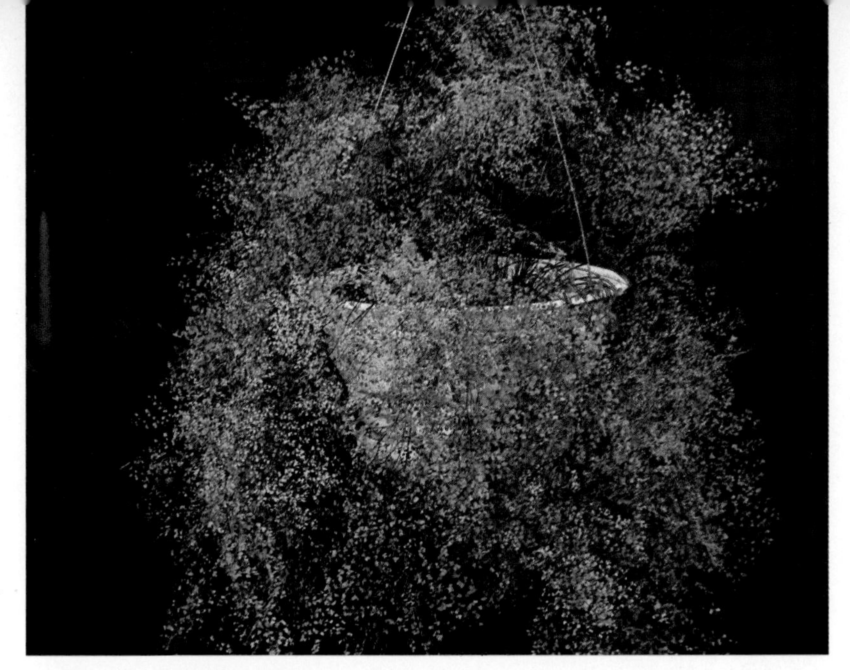

BABY'S-TEARS *Adiantum raddianum microphyllum*

The charm of this Maidenhair is its minute leaflets that rarely exceed ⅛" (.2cm) in width. High humidity.

DELTA MAIDENHAIR *Adiantum raddianum*

Filmy, pea-green fronds turn dark green when mature. Pinnae (leaflets) are wedge-shaped. Better as a greenhouse plant, but may be tried as a houseplant by providing extra humidity. Also known as *Adiantum cuneatum.*

MING MAIDENHAIR
Adiantum 'Ming Maidenhair'

Tight clustered mound of miniature fan-shaped leaflets, cool green.

Culture: Maidenhair Ferns

Temperature: Average climate, minimum at night 55°F (13°C); will tolerate temperatures of 45°F (7°C) for short periods. Optimum daytime temperature is 70°F (21°C).

Light: Low to medium; protect from direct sun. Good in unobstructed north light or in filtered light of an east window. Prefers humidity range of 40% to 60%. Curling and browning of new fronds may be an indication of too low humidity.

Watering: Keep soil uniformly moist but not wet. Avoid overwatering. In winter, allow soil to become slightly dry before rewatering, but do not allow plant to wilt.

Fertilizing: Only during active growth with dilute solution every 3 weeks or space applications every 4-6 months according to growth.

Soil: Add peatmoss, leafmold, or other humus to a houseplant mix. Sharp sand or perlite will add drainage. Most Maidenhair ferns like calcium-rich soils, so add dolomite limestone or oyster shells to the mix.

Special Considerations: Most Maidenhairs have a dormant period in winter, and it is advisable to rest them during this time. Cut off all old and discolored fronds to make way for new growth in spring. Reduce watering and set plant in a cool place. In spring, repot or top dress with fresh soil mixture. When small, these ferns are best used in terrariums.

Caution: Delta Maidenhair ferns can be damaged by some insect sprays. See Pests and Disease, page 76.

OCEAN SPRAY
Adiantum raddianum 'Ocean Spray'
Overlapping fronds give this compact fern a light, fluffy look. New growth is a delicate, light green.

FAN MAIDENHAIR *Adiantum tenerum 'Wrightii'*

Graceful, fan-shaped fronds will flow over the edge of a planter making this an excellent hanging or pedestal plant. New growth is tinged pink. A good choice for home cultivation.

GIANT MAIDENHAIR
Adiantum trapeziforme

Diamond-shaped fronds with finely serrated edges make this a unique and unusual Maidenhair. Although robust, it is best grown in a greenhouse unless the warm, moist conditions can be duplicated elsewhere.

PACIFIC MAID
Adiantum raddianum 'Pacific Maid'

Stacked one above the other, the leaflets (pinnae) are brillant yellow-green turning to a satiny, dark green when mature. Compact with a beautiful contrast of color, suitable for home conditions.

FIVE-FINGERED MAIDENHAIR
Adiantum pedatum

Dainty, excellent for northern areas. Stalks are a polished purplish black with palmlike, pea-green leaflets, height 12-24" (30-60cm). Attractive from early spring to fall. Deciduous, easy.

Culture: Loose, rich soil containing humus. Keep moist in filtered to deep shade. Eventually spreading but slow growth makes plant manageable for long time.

Uses: Foundation planting or in bed as accent with flowers, especially roses. Also, excellent in wild flower garden.

Bear's-Paw Fern

Aglaomorpha meyeniana

An epiphytic (tree-dwelling) fern with glossy, yellow-green fronds that emerge without a stalk from a plush, pawlike rhizome. The upper pinnae of a fertile (spore-producing) frond appear beaded and filmy. A rare and beautiful specimen for a hanging basket.

Culture: Bear's-Paw Fern

Temperature: Average to warm, night minimum not below 60°F (16°C).

Light: High light; bright diffused sunlight in winter; full to partial shade in summer. Needs high humidity.

Watering: Keep fairly moist.

Fertilizing: Very little needed; apply ¼ recommended strength every 4-6 months.

Soil: An osmunda fiber pot lined with shredded sphagnum and filled with peatmoss or other humus is ideal. Use sharp sand for good drainage.

Australian Tree Fern

Alsophila cooperi (australis)
(bot. Sphaeropteris cooperi)

A focal point of beauty for greenhouse or conservatory, this fern has broad, triangular fronds. The stem and young coiled leaves (croziers) are covered with coppery bristles. Eventually the stem forms a thick trunk resembling a tree and reaching a height of 8′ (2.4m). The trunk acts as a moisture chamber. It should be watered along with the soil because roots of the new fronds develop at the top of the trunk and grow down.

Culture: Australlan Tree Fern

Temperature: Warm climate; minimum at night 30°F (-1°C). Although frost and short periods below freezing are tolerated, foliage may die back.

Light: High light; bright diffused sunlight. Full sun in winter. Prefers high humidity.

Watering: Keep constantly moist in summer; barely moist in winter.

Fertilizing: Only during active growth with ¼ the recommended strength every 2 weeks.

Soil: Humus soil with addition of sphagnum moss, leafmold, sand or perlite. Top mulch of redwood chips ideal for retention of soil moisture.

Special Considerations: A challenge without the warm, moist conditions of a greenhouse, this plant is one of the easier of the tree ferns. To grow it in the North without a greenhouse, try this method; treat as an outdoor patio plant in a shaded spot from early spring to late fall. Winter under artificial or subdued light in a cool place (48-60°F/9-16°C). Cool temperatures and reduced light produce semidormancy, and the need for constant soil moisture and high humidity is reduced. Thus, the plant is easier to care for.

Maidenhair Spleenwort

Asplenium trichomanes

Small, spreading fern with dark green roundish leaflets. Excellent for northern areas. Height 4-6" (10-15cm). Evergreen.

Culture: Garden soil mixed with humus, and a small amount of horticultural limestone added at the time of planting. If planting in limestone rock, it is not necessary to add limestone to the soil. Create a pocket in a rock wall for fern and anchor it with rocks or gravel. Soil should be well drained with this positioning. Keep moist, if rainfall is inadequate. Fern tolerates some sunlight and dry periods.

Uses: Excellent in rock wall, rock garden, or as a low-growing border plant, especially around a fountain or small pond.

Bird's-Nest Ferns

Asplenium nidus

Above view, showing "nest" or erect rhizome of the Bird's-Nest Fern.

BIRD'S-NEST FERN
Asplenium nidus

Unfernlike in appearance, this plant has broad, polished elliptical fronds that are parrot-green in color. Egg-shaped new fronds uncurl from the center of the plant, which contains a dark brown, circular rhizome. The rhizome and the shape of the new fronds give this plant not only its botanical and common names, but also make it an interesting conversation piece. A good houseplant.

Culture: Bird's-Nest Fern

Temperature: Average to warm, minimum at night 62-65°F (17-19°C); if temperatures are too cold or too damp, brown spots will appear on the edges of leaves. This is not to be mistaken for the natural aging of the exterior leaves, which will eventually darken and die.

Light: High-medium; filtered sunlight to partial shade. Benefits from winter sun.

Watering: Keep uniformly moist but not wet. During the winter the plant should be kept moist to dry.

Fertilizing: Use ¼ recommended strength every 4-6 months to improve color of foliage.

Soil: This fern is epiphytic (tree-dwelling), and will benefit from soil rich in humus. Since roots are not extensive, use half-pot, or half-fill a standard pot with broken crockery before adding potting mixture.

Special Considerations: Humidity range 30-50%. Generally, tolerates drier air than majority of ferns. Give occasional sink shower with tepid water to clean dust off fronds. Fronds bruise easily; avoid handling.

LASAGNE FERN
Asplenium nidus 'Crispafolium' (Phyllitis scolopendrium 'Undulatum' in trade)

Glossy, bright green, pleated fronds make this an elegant specimen plant. Unlike its parent, *Asplenium nidus,* this fern presents a challenge to the grower. Greenhouse conditions, warmth and humidity, should be given.

Mother Ferns

MOTHER FERN

Asplenium daucifolium (viviparum)

Like a miniature evergreen, this fern appears with deep green, arching fronds. Like *A. bulbiferum*, new plants grow on the surface of mature fronds. The leaflets (pinnae) of these new ferns are of tear-drop shape with fine scalloping around the edges. A graceful fern of easy culture, this makes a welcome addition to a woodland terrarium. It does not exceed 10" (25cm) in home cultivation.

Culture: Mother Fern

Temperature: Adapted to wide range of houseplant temperatures, but not below 50°F (10°C).

Light: Low to medium; bright diffused sunlight to partial shade. Prefers humid air.

Watering: Keep uniformly moist but not wet.

Fertilizing: Apply ¼ recommended strength every 2 weeks during summer or space applications every 4-6 months.

Soil: Any houseplant mix with leafmold or other humus.

HEN-AND-CHICKENS FERN

Asplenium bulbiferum

Soft green, filmy carrotlike fronds characterize this fern that gives birth to miniature reproductions of itself. The small, round buds that appear on the mature fronds, sprout and form new ferns in an almost continuous process. Weigh fernlets down with moist soil or detach and plant in a terrarium-like container to provide humidity and prevent wilting. Two years are required for an adult fern grown from a fernlet. Good terrarium plant when young; accent plant of nearly 2' (60cm) when mature.

Lady Fern

Athyrium filix-femina

Feathery, tapering lancelike leaves to 18-36" (45-91cm) in length. This hardy fern is typically yellow-green. Foliage will brown slightly with maturity or if soil is not kept sufficiently moist. Deciduous, easy.

Culture: Loose, rich soil containing humus. Keep moist for lush appearance. Will tolerate direct morning or late afternoon sun. Spreads slowly.

Uses: Good for a foundation planting.

Note: There are many varieties of the hardy Lady Fern that are elegant with their crests and frills, and are just as easy to grow as the parent fern.

Japanese Painted Fern

Athyrium goeringianum

While mature foliage of most hardy ferns is limited to variations of green, this fern is a colorful departure. The fronds are wine-red with a central band of gray on green. They are usually 12" (30cm) in length, but may grow to 24" (60cm). New croziers (shoots) unfold throughout the summer. Deciduous.

Culture: Loose, rich soil containing humus. Keep moist to prolong beauty late into the season. Partial shade. Easy to grow and adaptable to cold winters.

Uses: Plant alone or among flowers; this fern with its quiet, colorful tones is a real acquisition and is sure to attract attention.

Rib Ferns

CRISPED RIB FERN
Blechnum brasiliense 'Crispum'

Slightly wavy, coarse green fronds display a reddish color when young.

Culture: Rib Ferns

Temperature: Average to warm climate; minimum at night not below 60°F (16°C).

Light: Medium light; filtered sunlight to partial shade.

Watering: Keep uniformly moist but not wet. Soak occasionally in a pail of water to saturate roots and prevent drying out.

Fertilizing: Every 4-6 months using ¼ recommended strength.

Soil: A well-drained, humus houseplant mix made slightly basic (alkaline) by addition of limestone or oyster shells.

RIB FERN
Blechnum gibbum

This fern has a tree-fern habit and develops a "trunk" with age. It is medium-sized and matures slowly. Fronds are coarse and deeply cut, the pinnae slightly wavy. An attractive accent plant.

Walking Fern

Camptosorus rhizophyllus

The unusual and suggestive name of this hardy fern intrigues everyone. The tapering fronds, 4-12" (10-30cm) long literally appear to be walking over their favorite habitat of rocks and ledges. Besides the conventional method of reproduction by spores, this fern literally "walks itself into existence." Wherever the tip of the frond rests, new roots are formed, and a new independent plant results. Evergreen.

Culture: Garden soil to which horticultural lime or crushed limestone has been added. Capitalize on the natural pockets of existing stones or create them between rocks. Keep moist for a few days after planting; once established the plant can endure short, dry periods. Place in shade to semi-shade. This fern is prey to slugs. Use a commercial bait to control them.

Uses: Excellent miniature, low-growing fern for rock gardens and ledges.

Rattlesnake Fern

Botrychium virginianum

Belonging to a genus known as the Grape Ferns, probably due to the arrangement of the spores in clusters like grapes. These spores are carried above a single, yellow-green, triangular frond and fruit in early summer, but are usually sterile (non-spore bearing) in their first year. Uniquely, the fern has no croziers, but instead develops its leaf underground. One of the earliest to appear in spring, the leaf remains green all summer. Deciduous.

Culture: Rich woodland-type soil. Keep moist for best results. Full to open shade.

Uses: Best as a groundcover plant, especially under trees. Will give a natural, woodsy appearance to landscape.

Mexican Tree Fern

Cibotium schiedei

Gracefully arching fronds are Nile green and lacy. The trunk is covered with yellowish brown to blackish brown hairs. Plant is often sold as 4-6′ (1.2-1.8m) trunks without fronds. Spectacular, exotic foliage plant for home or conservatory, remarkably adaptable to indoor culture even under dry indoor conditions in cold climates.

Culture: Mexican Tree Fern

Temperature: Average to warm; night minimum 50°F (10°C).

Watering: Constantly moist but not wet. Tolerates occasional drying out and a drier, less humid atmosphere.

Light: Medium light; diffused sunlight to partial shade. Full shade in summer.

Fertilizing: Only during active growth with ¼ the recommended strength every 2 weeks.

Soil: Spongy and loose mixture composed of shredded sphagnum, leafmold or other humus in a houseplant mix. Use sand or perlite for good drainage. Top mulch with redwood chips.

Special Considerations: Found growing naturally in lava rock, and it is possible to grow plant for a number of years in water with rocks or pebbles to hold it erect. Add charcoal to water for purity, and fertilize year-round if water planting is used.

Holly Ferns

HOLLY FERN

Cyrtomium falcatum 'Rochfordianum'

This cultivar has a stronger hollylike appearance than *C. falcatum*. Metallic green, the pinnae (leaflets) are broad and finely toothed with distinctly pointed tips. Fronds are tightly curled when they first appear, and unfold slowly to their characteristic shape; a robust and durable houseplant.

Culture: Holly Fern

Temperature: Generally cool, but adaptable to normal range of household temperatures.

Light: Medium to low light.

Watering: Keep uniformly moist, but not wet; will tolerate short periods on the dry side.

Fertilizing: Every 2 weeks during active growth in late spring and early summer with weak solution.

Soil: Any general houseplant mix to which sand or perlite has been added to insure good drainage.

JAPANESE HOLLY FERN

Cyrtomium falcatum 'Butterfieldii'

Similar to the cultivar 'Rochfordianum' with slightly smaller, shiny, green, serrated leaf margins. Does well in low light and humidity.

FISHTAIL FERN
Cyrtomium falcatum

A medium-sized evergreen fern of easy culture that will tolerate drafts, low light, and the drier atmosphere of most homes. It resembles Christmas Holly with glossy, dark green, spiked fronds. A very durable houseplant.

Holly Ferns of the Cyrtomium Family grow equally well outdoors in milder climates, and will tolerate frost if care is taken to mulch the roots.

Culture: Fishtail Fern

Temperature: Generally cool, but adaptable to average household temperatures, 65-70°F (19-21°C).

Light: Medium to low light. Will thrive in shade without direct sun.

Watering: Keep uniformly moist, but not wet. Will tolerate short periods of dryness without collapse.

Fertilizing: Every 2 weeks during active growth in late spring and early summer with ¼ recommended strength.

Soil: Any general houseplant mix, loose, with sand or perlite to insure good drainage.

Rabbit's-Foot Ferns

BALL FERN
Davallia mariesii

Small, epiphytic fern with shiny, green, triangular fronds supported on wiry stems. Davallias are prized for their exposed furlike rhizomes. Rabbit's-Foot, Hare's-Foot, Squirrel's-Foot are common names for these ferns.

RABBIT'S-FOOT FERN

Davallia fejeensis 'Plumosa'

The most delicate and finely cut of the davallias. To propagate, break off a "foot" (rhizome) and place on moist soil. Cover with plastic or inverted jar for humidity, and soon a new fern will emerge. An easy, attractive plant.

Culture: Davallia Ferns

Temperature: Average to warm; not below 60°F (16°C); deciduous below this temperature.

Light: Adaptable from high to low-light intensities; shade from direct sun. Prefers humid air.

Watering: Keep soil uniformly moist but not soggy wet.

Fertilizing: Apply ¼ recommended strength several times during growing season for proliferation of rhizome.

Soil: Porous soil to which a combination of shredded sphagnum, peatmoss, and humus has been added.

Special Consideration: For easy care and good drainage, plant fern in carved osmunda bark container or wire basket. An osmunda bark container is best. Otherwise, be sure to drape rhizomes over sides of pot and take care not to bury them in the soil.

SQUIRREL'S-FOOT FERN

Davallia trichomanoides

Medium-sized fern with arching, graceful triangular fronds, spectacular as a hanging-basket plant. Rust-brown rhizomes enhance its decorative quality. Once established, an easy and rewarding fern.

Hay-Scented Fern

Dennstaedtia punctilobula

Delicate, tapering, pale yellow-green fronds, 20-32" (50-80cm) long, are sweet-scented as they dry or when crushed. It spreads rapidly, and is so adaptable that it is often considered a nuisance. Deciduous.

Culture: Tolerates a variety of soils as well as prolonged wet or dry periods. Accepts deep shade to full sunlight.

Uses: Makes an excellent lush, dense groundcover for slopes or otherwise scrubby, barren land.

New Zealand Tree Fern *Dicksonia squarrosa*

Coarse to the touch, this fern has yellow-green triangular fronds. At the base, the leaf stalk is covered with shaggy, orange-brown hairs. An easy tree fern of medium size, it is decorative as an accent for home, greenhouse, fernery, or outside as a tub plant for a shaded patio.

Culture: New Zealand Tree Fern

Temperature: Average to cool climate; can tolerate temperatures as low as 20°F (-7°C).

Light: High-medium light; diffused sunlight to partial shade. Prefers high humidity.

Watering: Keep moist but not wet. Mist base or trunk as well as foliage.

Fertilizing: Only during active growth with ¼ recommended strength every 2 weeks.

Soil: Add peatmoss or leafmold to an equal volume of general houseplant mix.

Tree Maidenhair Fern

Didymochlaena truncatula (lunulata)

Closely set, somewhat rectangular leaflets are colorful, ranging from maroon to bronzy, metallic green. Although the appearance of the fern is leathery, the leaflets have a papery quality to the touch. This fern grows 1-3' (30-91cm), and can be considered botanically unique or rare, since there is only one species—the one you see pictured, no other varieties or cultivars.

Culture: Tree Maidenhair Fern

Temperature: Average to warm climate, night minimum 60°F (16°C).

Light: Medium light, either indirect sunlight or partial shade. Prefers humid air.

Watering: Keep soil uniformly moist but not wet.

Fertilizing: Apply dilute strength during active growing period, April through October, every 2 weeks.

Soil: Add peatmoss or leafmold to an equal volume of general houseplant mix. Provide good drainage.

Wood Ferns

MALE FERN
Dryopteris filix-mas

The male counterpart to the Lady Fern, this plant has equally fancy plumage. All leaflets, with exception of those at the tip, are feathered and crested. The over-all shape of the frond is lance-like with an erect stature of almost 36" (91cm) and a breadth of 9" (23cm). Deciduous.

Culture: Make soil woodsy by addition of humus, commercial peatmoss, leafmold, or use a compost of well-rotted leaves. Keep moist. Full shade to semishade.

Uses: Show fern to use in prized flower bed or as a backdrop for lower growing plants.

JAPANESE SHIELD FERN
Dryopteris erythrosora

Beautiful low-growing hardy fern with copper-red fronds 10" (25cm) long, that change to a leathery, deep green when mature.

Culture: Good garden soil with humus, peatmoss or leafmold added. Moist, but will tolerate dry periods. Enjoys shade, but accepts some sunlight.

Uses: Good rock-garden plant.

TOOTHED WOOD FERN
Dryopteris austriaca spinulosa

This species abounds in New England woods, and is sometimes called the Florist's Fern because of its commercial use. It is finely cut with clustering, lance-shaped yellow-green fronds that can grow to 30" (75cm). Nearly evergreen.

Culture: Add peatmoss, or leafmold to create slightly acid soil. Mulch and keep moist, especially during hotter summer months. Full to semishade. Nonspreading.

Uses: Excellent as border around base of trees or with acid-loving shrubs like azaleas or rhododendrons. For a woodland scene, plant in the hollow of a tree stump.

Hand Fern

Doryopteris pedata

A plant that shatters the popular idea of a fern is this handsome one with polished chestnut-brown stems and broad, maple-like leaves.

Culture: Hand Fern

Temperature: Average to warm; minimum at night not below 60°F (16°C)

Light: Medium light to partial shade. Prefers humid air.

Watering: Keep uniformly moist but not wet.

Fertilizing: Apply ¼ recommended strength every 4-6 months.

Soil: Humus soil that permits good drainage.

Japanese Climbing Fern

Lygodium japonicum

An upward climbing yellow-green fern that looks more like a vine than a fern. The plant adapts well to a trellis, or grow so fronds cascade from a hanging planter. Try one on a fern "log". It will climb to cover an 8′ (2.4m) post.

Culture: Japanese Climbing Fern

Temperature: Average to cool; minimum night 50°F (10°C).

Light: High-medium light; filtered sunlight in summer. Good choice for a window garden. Prefers humid air.

Watering: Keep uniformly moist but not wet.

Fertilizing: Apply ¼ recommended strength every 4-6 months.

Soil: Any houseplant mix suitable so long as there is good drainage.

Boston Ferns

DWARF BOSTON FERN
Nephrolepis exaltata 'Compacta'

Attractive, dwarf species with dense, shrubby habit. Fronds are more upright than pendant. Easy plant for home cultivation.

SWORD FERN
Nephrolepis exaltata

The original fern from which so many variants have been produced. Leaflets are plain, with the entire frond long and lancelike. Fronds may grow to 4′ (1.2m) in home cultivation. Plant produces spores, and plants so propagated will be bushier that those that have been divided. Stands considerable sunlight, even outdoors in temperate coastal areas.

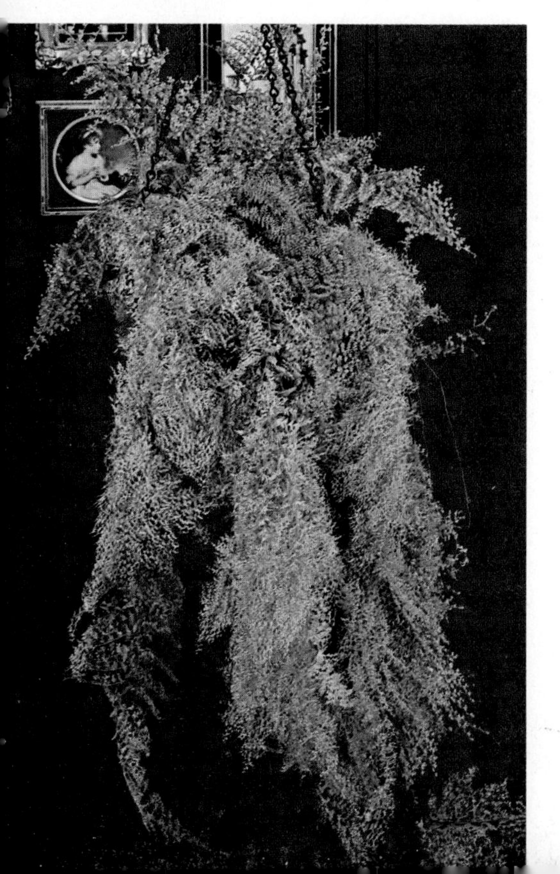

BOSTON FERN

Nephrolepis exaltata 'Bostoniensis'

An old-fashioned favorite with long (up to 10'/3m), tapering yellow-green fronds. This is a durable fern for home or office that will tolerate air conditioning but not drafts.

Culture: Boston Fern

Temperature: Average to cool, night minimum 50-55°F (10-13°C).

Light: Bright diffused sunlight to partial shade. Benefits from winter sun. Prefers humid air.

Watering: Keep soil moist but not wet. Large plants can be watered by immersing pot in tub of water for approximately 1 hour.

Fertilizing: Apply ¼ the recommended strength every 4-6 months.

Soil: A mixture of houseplant soil, sand or perlite, and humus is adequate. Insure good drainage.

Special Considerations: Although some Boston Ferns reproduce by spores, they all spread by threadlike runners called stolons. These appear at the base of the plant, and can be inserted into the soil to produce offshoots or to enhance bushiness. They can also be trimmed off with no harm to the plant. Best if potted in spring when new growth starts.

VERONA LACE FERN

Nephrolepis exaltata 'Verona'

Dwarf and delicate, yellow-green lacy fronds grow long and trailing. Do not mist foliage but provide good humidity.

◀ **BOSTON PETTICOAT**
Nephrolepis exaltata 'Petticoat'

Tufted leaf margins add interest to the long, pendant fronds.

MINI-RUFFLE FERN ▲
Nephrolepis exaltata 'Mini-Ruffle'

Petite fern with compact, lacy, ruffled fronds that do not exceed 2" (5cm). Useful for terrariums and dish gardens.

◀ **FISHTAIL FERN**
Nephrolepis biserrata 'Furcans'

This fern derives its common name from the widely spaced, deep green pinnae which are forked and resemble a fishtail. Best in a greenhouse or with comparable high humidity.

CRISPED FEATHER FERN ▶
Nephrolepis exaltata 'Hillsii'

Vigorous, fronds are long with crisped and wavy leaflets.

FLORIST FANTASY ▶

Nephrolepis exaltata 'Erect'
(Pat. Pend.)

Tightly curled and undulating leaflets flank an erect, silver-scaled midrib to make this a prize variety of Boston fern.

◀ SHADOW LACE FERN

Nephrolepis exaltata 'Shadow Lace'

Brushy habit, leaflets are tightly clustered and overlapping, giving it a lacy look.

Culture: Boston Fern

Temperature: Average to cool, night minimum 50-55°F (10-13°C).

Light: Bright diffused sunlight to partial shade. Benefits from winter sun. Prefers humid air.

Watering: Keep soil moist but not wet. Large plants can be watered by immersing pot in tub of water for approximately 1 hour.

Fertilizing: Apply ¼ the recommended strength every 4-6 months.

Soil: A mixture of houseplant soil, sand or perlite, and humus is adequate. Insure good drainage.

Special Considerations: Although some Boston Ferns reproduce by spores, they all spread by threadlike runners called stolons. These appear at the base of the plant, and can be inserted into the soil to produce off shoots or to enhance bushiness. They can also be trimmed off with no harm to the plant. Best if potted in spring when new growth starts.

FLUFFY RUFFLES

Nephrolepis exaltata 'Fluffy Ruffles'

A miniature variety similar to 'Florida Ruffle'. Ruffled fronds grow to 12" (30cm).

▼ FLORIDA RUFFLE FERN
Nephrolepis exaltata 'Florida Ruffle'

Stiff, upright fronds are dense and lacy-looking. Although compact, this is not a dwarf, fronds may grow to 24" (60cm).

TEDDY JR. FERN ▼
Nephrolepis exaltata 'Teddy Jr.'

Robust, the yellow-green fronds are slightly wavy on this compact, dwarf fern.

◂ DWARF WHITMANII FERN

Nephrolepis cordifolia 'Plumosa'

Arching, long stiff fronds are narrow with tufted leaf margins. This plant bears spores on the underside of the leaves.

▾ PYGMY SWORD FERN

Nephrolepis duffii

An unusual fern of compact habit. The fronds are clusters of ruffled, button-shaped leaflets.

ROOSEVELT FERN ▾

Nephrolepis exaltata 'Rooseveltii'

Large, simply elegant fern with subtle undulation in the leaflets.

SILVER BALLS FERN
Nephrolepis exaltata 'Silver Balls'

Long, erect fronds are crinkled with a metallic silver luster on new growth.

Sensitive Fern *Onoclea sensibilis*

The individual leaflets, 12-30" (30-75cm) long, give more the appearance of an oak than a fern. Leathery, yellow-green fronds are contrasted with the fertile fronds, which appear in June or July, and are not leaves, but spore-bearing, bead-shaped clusters. If ever a fern could be called a weed, it would be this one because of its tendency to spread rapidly. Leaves are not sensitive to the touch, but the name reflects sensitivity to early frosts. Deciduous.

Culture: Ordinary garden soil that alternates between wet and dry. Semishade, but will tolerate considerable sunlight if soil moisture is abundant.

Uses: Excellent cover for damp locations without heavy shade. An especially good outdoor fern for southern California.

Osmunda Ferns

INTERRUPTED FERN
Osmunda claytoniana

Tall, golden-green fronds to 4′ (1.2m), erect and curving outward from a tight rosette. Cottonball-like shoots (croziers) appear early in spring. The upward path of the leaflets on the stalk is interrupted by "flowering clusters," which are the spore-bearing segments. Deciduous.

Culture: Soil should be made acid by addition of peat or leafmold. Swamp-muck, if it can be obtained, is excellent. Soil should be continuously damp, if large plants are desired. Less ideal conditions will result in somewhat smaller plants. Place in shaded area.

Uses: Fern does not spread rapidly. Best as an accent or backdrop.

ROYAL FERN
Osmunda regalis spectabilis

Unique fern with leaflets like those of a locust tree. Small, spherical flowerlike clusters containing spores (fruit) at the top contribute to the non-fern appearance. Leaflets are tinged red when young, changing to deep green at maturity. Overall height of plant is 3-4′ (.9-1.2m). Deciduous.

Culture: Acid soil containing generous amounts of peat or leafmold. Swamp muck, if available, is a good additive. Adaptable, semishade to full sunlight. Tolerance to increasing levels of sunlight is dependent on quality and moisture of the soil.

Uses: Good along a brook in a woodland garden, or at an entrance to deep woods. Use as transitional planting between cultivated garden area and woodland.

Cliffbrake Ferns

BUTTON FERN
Pellaea rotundifolia

A low, spreading fern with button-shaped leaflets (pinnae) makes a striking plant for a low table or hanging basket. A choice easy houseplant. Also good for terrariums and groundcover.

Culture: Button Fern

Temperature: Cool to average; semihardy, will tolerate short periods below freezing. Can be planted outdoors in milder climates.

Light: High light; growth enhanced by winter sun. Otherwise, bright, diffused sunlight. Prefers humid air.

Watering: Pellaeas are xerophytic ferns. They are native to dry regions, and have special features that permit revival even if plants are allowed to dry out. Therefore, keep moist but allow soil to dry slightly between waterings.

Fertilizing: Growth slows as weather cools; fertilize with ¼ recommended strength every 2 weeks only during active growth.

Soil: Loose, humus soil with ample crocks for drainage. Not necessary to add limestone for houseplant cultivation.

Special Considerations: Fern does not need deep pot. Diameter of pot should be wide or plan on repotting at frequent intervals to accomodate fast-growing and spreading rhizome.

GREEN CLIFFBRAKE
Pellaea viridis

A wiry, delicate fern with medium-green lancelike leaves that are almost paper thin. Bushy habit; fronds grow rapidly to 30" (75cm).

Hart's Tongue Ferns

Phyllitis scolopendrium

A vigorous fern with strap-like wavy fronds, said to resemble the tongue of an animal. Foliage is a beautiful, glossy yellow-green with wavy edges. Length of fronds is from 6 to 18" (15-45cm) and entire plant has a dense, bushy habit. Dusty brown spores are often seen on the underside of fronds. They form a herringbone pattern and should not be mistaken for a pest.

Culture: Hart's-Tongue Fern

Temperature: Average to cool; fern is hardy and will tolerate long periods below freezing if outdoor planting is desired.

Light: Low light; bright diffused sunlight to partial shade. Benefits from winter sun. Will tolerate some sunlight outdoors.

Watering: Moist, but allow soil to dry out slightly between waterings. Avoid over-watering as plant is susceptible to root rot.

Fertilizing: Fern is active year round; fertilize every 2-4 weeks with ¼ recommended strength.

Soil: Any houseplant mix, enriched and made slightly basic by adding dolomite limestone, oyster shells, or crushed limestone rock. When planting outside, provide a basic soil composition by adding horticultural lime, crushed limestone rock or oyster shells to a prepared bed.

Special Considerations: Fern may be propagated by inserting leaf stalk in a moist medium. Excellent for rock garden, terraced stone walk and as border fern in a shaded garden.

An excellent outdoor fern for northern areas, the Hart's Tongue is exceptionally easy to care for. Not suited to boggy areas; be sure to provide good drainage.

Staghorn Ferns

COMMON STAGHORN FERN

Platycerium bifurcatum

Slow-growing epiphytes (air feeding) with fuzzy antler-shaped fronds; Platyceriums are curios of the plant world. They are easy plants with this species being the choice for home conditions. The plant has a parchment-like shield that can be attached to a slab of osmunda bark for hanging; also suitable for pot culture.

Culture: Common Staghorn Fern

Temperature: Average to cool; night minimum 50-55°F (10-13°C). Tolerates short periods below freezing.

Light: High light. Sun in winter, from mid-March onward filtered sun to full shade. Prefers moderately humid air.

Watering: Keep moist but allow soil to dry slightly between waterings. If plant is hung on a slab, place it biweekly in a bowl or sink filled with tepid water. Soak for 15 minutes or more and drain before rehanging. (The shield's purpose is to store moisture.)

Fertilizing: Apply ¼ recommended strength only during active growing period. Use only liquid fertilizer to avoid build-up in between layers of shield.

Soil: Pot — fibrous mixture containing sphagnum and peatmoss with sand for grit. Slab — place 1-2" (2.5-5cm) of sphagnum moss behind shield before attaching it.

REGINA WILHELMINA STAGHORN
Platycerium bifurcatum 'Netherlands'

Distinguished from Common Staghorn Fern by a darker green that contrasts more readily with the grayish white felt covering that is typical on fronds of Staghorn ferns. Fronds are also broader without clear cuts or divisions. They arise in a circular, starlike fashion from a cushion (basal frond) of pale green. One of the best choices for houseplant culture.

ELK'S-HORN FERN
Platycerium hillii

Another staghorn suitable for home cultivation, the fan-like antler fronds make a striking wall decoration when mounted on a slab of rough wood.

Hare's-Foot Ferns

RHIZOME OF HARE'S FOOT FERN

Close-up of the colorful rhizome (foot system) of this fern. Measuring 1" (2.5cm) in diameter, rhizome climbs over and around the container as the plant grows.

GOLDEN POLYPODY
Polypodium aureum

A large specimen plant with lancelike, deeply cut leaves that can span 10" (25cm) across and grow to a length of 36" (91cm). Color is a soft blue-green and 1" (2.5cm) diameter rhizome is orange-brown.

◀ **HARE'S-FOOT FERN**
Polypodium aureum 'Undulatum'

Arching, deeply-cut fronds undulate, giving this fern a bold look. The fronds may grow over 3' (.9m) in length and 6-8" (15-20cm) wide. An easy fern to grow.

HARE'S-FOOT FERN

Polypodium aureum areolatum (glaucum)

Blue-green, 15-20" (37-50cm) fronds produced from a stout rusty brown rhizome. The durable fronds have a glaucous or frosted underside. An easy fern to grow.

Culture: Hare's Foot Ferns

Temperature: Average to warm climate; tolerates minimum night temperature of 50-55°F (10-13°C).

Light: Medium light; bright diffused sunlight to partial shade. Benefits from winter sun. Try for humid atmosphere, but will tolerate drier home conditions.

Watering: Keep uniformly moist, but not wet. Will tolerate drier soil in winter provided rhizome is misted daily with tepid water.

Fertilizing: To enhance growth of root system (rhizomes), fertilize with ¼ recommended strength every 3 weeks. For slower growth, apply every 4-6 weeks.

Soil: Any good houseplant mix enriched with leafmold, humus, or peatmoss. Must drain well.

Polypody Ferns

WART FERN
Polypodium scolopendria

Deeply lobed, dark emerald-green fronds that are stiff and leathery make this plant a striking contrast to filmy ferns. Rhizomes are most unusual, appearing waxy and wiry; green with blackish bristles. A bold accent best displayed as a hanging-basket plant.

Culture: Polypody Ferns

Temperature: Average to warm climate; minimum night 60°F (16°C).

Light: Medium light; diffused sunlight to partial shade. Prefers humid air; but can tolerate drier atmosphere.

Watering: Keep uniformly moist but not wet.

Fertilizing: Apply ¼ recommended strength every 3 weeks; for slower growth every 4-6 months.

Soil: A well-drained plant mix containing equal parts peatmoss, shredded sphagnum, sharp sand.

CLIMBING BIRD'S-NEST FERN ▼
Polypodium punctatum 'Grandiceps'

Erect, yellow-green forked and crested fronds characterize this slow-growing fern. A specimen plant of easy culture that withstands drier home conditions.

RIBBON FERN ▲
Polypodium angustifolium

Lacquered green and bronze straplike leathery fronds grow to 2' (60cm). Excellent for hanging basket.

WALL POLYPODY
Polypodium vulgare

Fronds are blunt, deeply cut, lancelike and somewhat leathery. Yellow-green and low-growing, the maximum height of 10" (25cm) is seldom achieved in cultivation. The fern is evergreen but does undergo a transition as the weather gets cold. The leaf will change to greenish brown, rusty brown and be marked with splashes of black. Curling is also evident. These weathered leaves persist even after new leaves appear. Easy.

Culture: Very tolerant fern. Grows in both alkaline and acid soil. Accepts dry periods. Will grow in semi-shade to full shade.

Uses: Plant among rocks and boulders or as a border plant along path or in flower bed.

RESURRECTION FERN
Polypodium polypodioides

A small evergreen fern with the remarkable ability to fold up its fronds to conserve moisture during dry spells; it unfolds them when rainy weather returns. To duplicate this process simply place fern in water and wait for it to revive.

Culture: Resurrection Fern

Temperature: Average household temperatures; outside in milder climates will tolerate short periods below freezing.

Light: Diffused sunlight to partial shade. Prefers humid air.

Watering: Moist on wet side. Never allow plant to dry out or dormancy will be triggered.

Fertilizing: None needed.

Soil: Requires highly acid soil for best growth. Mix equal volume of peatmoss or leafmold with houseplant mix to create acidity. Good drainage essential.

Shield Ferns

CHRISTMAS FERN
Polystichum acrostichoides

Dagger-shaped fronds are truly ornamental with coloration varying from polished deep green to blue-green. Stalks are covered with rust-brown bristles. Croziers are tightly coiled and hoop-shaped. Normal height 1-2′ (30-60cm) but occasionally will grow to 3′ (90cm). Evergreen, easy.

Culture: Garden soil enriched with peat or leafmold. Full to semishade. Will tolerate considerable sunlight provided soil is kept very moist.

Uses: Accent plant, moderately low growing, spreads so slowly that containment is not a problem.

HEDGE FERN
Polystichum setiferum

Brown wooly leaf stalk, and coarse, but filigreed fronds.

Culture: Hedge Fern

Temperature: Cool climate; hardy outdoors withstanding long periods below freezing.

Light: Medium light; filtered sunlight to partial shade

Watering: Moist but allow to dry slightly between waterings. Tolerates occasional drying out without collapse of foliage.

Fertilizing: Only during active growth, using ¼ recommended strength.

Soil: Any general houseplant mix to which extra humus has been added. Provide good drainage.

BRISTLE FERN
Polystichum setosum

Low, spreading fern with brown, hairy stalks and glossy, deep green, lacy fronds that end in a bristle. Croziers are unique as the tip is soft, silver white and looks like the tail of a rabbit or deer.

Culture: Bristle Fern

Temperature: Average to cool climate, will tolerate short periods below freezing.

Light: High-medium light; bright diffused sunlight to partial shade.

Watering: Keep soil moist but allow to dry slightly between waterings.

Fertilizing: Apply ¼ recommended strength every 2 weeks during active growth.

Soil: Add peatmoss, leafmold to an equal volume of packaged houseplant mix. Provide good drainage with addition of sharp sand or perlite.

DWARF LEATHER-LEAF FERN
Polystichum tsus-simense

A small, compact fern ideal for terrariums and dish gardens. This fern is deep green with sharply cut, triangular fronds. Larger specimens are excellent for basket and window culture.

Culture: Dwarf Leather-Leaf Fern

Temperature: Generally cool, but adaptable to warmth and humidity of closed terrariums, 60-70°F (16-21°C) day to 32°F night (0°C).

Light: Medium to low light. Will thrive in semishade without direct sun.

Watering: Keep soil uniformly moist, but not wet. Will tolerate short periods of dryness.

Fertilizing: Apply ¼ recommended strength every 4-6 months.

Soil: Any good houseplant mix with added humus. Supply good drainage.

Brake Ferns

BRACKEN OR BRAKE
Pteridium aquilinum latiusculum

A well-known vigorous and striking fern of hillsides, fields, and open woods, seldom used as an ornamental except where its rapidly spreading character is desired. Usually grows solitarily in dense fern fields, is 1-2½' (30-75cm) high, cut in three divisions giving it a winged effect. The handsome leaf is coarse and leathery and turns brown in autumn. Deciduous.

Culture: Although once established, this is easy to grow, successful transplanting is not easy. Several forays to the woods may be necessary to get it to grow as a garden plant. It will tolerate considerable sun and dryness. Ordinary garden soil is sufficient and it is valuable for sandy soils.

Uses: A good spreading fern for hillsides and slopes to aid in prevention of soil erosion. Also, good as a transitional planting between home garden and a woodland tract.

Table Ferns

VARIEGATED TABLE FERN

Pteris cretica 'Albo-lineata'

A handsome fern with a broad band of creamy-white down the center of each slightly wavy leaflet.

CRETAN BRAKE ▼

Pteris cretica 'Gautheri'

Similar to the Lacy Table Fern, but with broad leaflets. Nicely compact and easy for terrariums, dish gardens, and planters.

Culture: Table Ferns

Temperature: Average; night minimum 55°F (13°C).

Light: Partial shade to bright diffused sunlight. Benefits from winter sun. Does well in low humidity.

Watering: Keep uniformly moist but not wet.

Fertilizing: Apply ¼ recommended strength every 2 weeks during active growth or elect 4-6 month schedule for slower, but steady growth.

Soil: Tolerates any general houseplant mix. Can be enriched with peatmoss or other humus. Provide good drainage.

◀ **PARKER TABLE FERN**
Pteris cretica 'Parkeri'

Small, compact plant similar to 'Major', only difference being that leaflets are broader and finely toothed.

LACY TABLE FERN ▼
Pteris cretica 'Rivertoniana'

Light green leaves are feathery like those of carrots. Dependable plant for pots or terrariums.

SILVERLACE FERN

Pteris quadriaurita 'Argyraea'

Large, blue-green fronds banded silver down the center. Lower leaflets are winged or butterfly-shaped. A beautiful fern to group with other ferns or to contrast with foliage or flowering plants.

Culture: Table Ferns

Temperature: Average; night minimum 55°F (13°C).

Light: Partial shade to bright diffused sunlight. Benefits from winter sun. Does well in low humidity.

Watering: Keep uniformly moist but not wet.

Fertilizing: Apply ¼ recommended strength every 2 weeks during active growth or elect 4-6 month schedule for slower, but steady growth.

Soil: Tolerates any general houseplant mix. Can be enriched with peatmoss or other humus. Provide good drainage.

SKELETON TABLE FERN

Pteris cretica 'Wimsettii'

Leaf segments are serrated with tasseled or forked tips. Excellent in containers of all kinds.

FAN TABLE FERN
Pteris cretica 'Wilsonii'

Crisp, bright green, fan-shaped fronds form dense crests at tips, resembling sprigs of parsley. A real beauty for table or terrarium.

LADDER BRAKE FERN *Pteris vittata*

Fast growing, dark green, graceful fern with arching fronds to 30" (75cm). General appearance is somewhat plumelike. Easily grown. Stands considerable light. Full morning sun or late afternoon sun outdoors.

AUSTRALIAN BRAKE FERN
Pteris tremula

Soft, yellow-green fronds with serrated edges grow 18-24" (45-60cm) long. One of the largest brakes. Although fronds are short-lived, fern is prolific with many new croziers (shoots) waiting to unfold. Remove old fronds to improve appearance and assist new growth. Tolerates winter sun and a drier atmosphere.

◂CHINESE BRAKE
Pteris multifida 'Variegata'

Elegant variety with long, ribbon-like leaflets beautifully banded in white and highly crested at the tips.

CRESTED BRAKE ▸
Pteris cretica 'Roweii'

Nothing more than a frilly, crested 'Parkeri', but a pleasing difference.

Culture: Table Ferns

Temperature: Average; night minimum 55°F (13°C).

Light: Partial shade to bright diffused sunlight. Benefits from winter sun. Does well in low humidity.

Watering: Keep uniformly moist but not wet.

Fertilizing: Apply ¼ recommended strength every 2 weeks during active growth or elect 4-6 month schedule for slower, but steady growth.

Soil: Tolerates any general houseplant mix. Can be enriched with peatmoss or other humus. Provide good drainage.

◀ **ORIENTAL BRAKE FERN**
Pteris cretica 'Major'

Erect, umbrella-like leaflets are a delicate, light green.

Pteris cretica 'Mayii' ▶

Almost identical to Variegated Table Fern, although smaller. Each leaflet has a center band of white. The narrower segments are forked and crested at the tip.

◀ **SILVER LEAF FERN**
Pteris ensiformis 'Victoriae'

Small fern ideal for dish gardens or terrariums. Leaflets are banded in silver and edged in green.

Leather Fern

Rumohra adiantiformis

The fronds look familiar as they are commonly used by florists with cut flowers. The dark green coarse fronds are long lasting. With its attractive bushy habit, it is a durable houseplant.

Culture: Leather Fern

Temperature: Average to cool, night minimum 50°F (10°C). Tolerates short periods below freezing in mild climates.

Light: Medium light; benefits from winter sunlight, shade from direct sun in summer. Tolerates low humidity.

Watering: Keep moist but allow soil to dry slightly between waterings, especially during winter.

Fertilizing: Every 2-4 weeks during active growth. Use ¼ the recommended strength.

Soil: Any houseplant mix with leafmold or other humus that permits good drainage.

Special Considerations: Excellent under-bench plant for greenhouses. Plant directly in ground for easy maintenance.

Felt Fern

Pyrrosia lingua

A small creeping epiphyte with thick, leathery fronds, ideal for pots or baskets. Exotic looking, yet easy to grow.

Culture: Felt Fern

Temperature: Average to cool climate, night minimum 50°F (10°C).

Light: Medium light; diffused sunlight or partial shade. Does well in low humidity, if care is taken to mist biweekly.

Watering: Keep soil just moist, not wet. To prevent root rot, avoid overwatering.

Fertilizing: Apply ¼ recommended strength every 4-6 months.

Soil: Grow in a wire basket or other pot allowing good drainage. Line basket or fill pot with sphagnum moss, leafmold, and sharp sand.

Climbing Fern

Stenochlaena palustris

Epiphytic fern with shiny, crisp-to-touch leaflets that are finely serrated at margins. New fronds are copper colored. Stout, pale green, creeping rhizome attaches easily to osmunda bark slabs for display. May be mounted on natural-looking logs or driftwood. Fill pockets in log with soil before planting.

Culture: Climbing Fern

Temperature: Average, not below 50°F (10°C).

Light: High light; filtered or diffused sunlight. Greenhouse atmosphere needed with high humidity to insure success.

Watering: Keep plant just moist, but not wet. Some dryness tolerated.

Fertilizing: Apply ¼ recommended strength every 4-6 months.

Soil: Humusy, containing mixture of sphagnum moss, peatmoss or leafmold. Provide good drainage.

Selaginellas

Fern allies are plants that are on the same evolutionary scale as ferns, as are the selaginellas or spike mosses and reed-like plants known as horse tails. Like ferns, they produce spores, but not the large leaves or fronds characteristic of ferns. Instead, they have tiny leaves on flat branches, which have a soft mossy or fern-like appearance.

Most selaginellas thrive in a humid greenhouse, but also in terrariums. Creeping forms, such as club mosses, make excellent groundcover there, putting out roots from numerous points on the branches.

Selaginellas are not difficult to keep in a terrarium. They require very little light; they also grow well suspended 1 to 16 inches below fluorescent tubes. Usually they do not require fertilization, but a solution of fish emulsion every 4 to 6 months can be beneficial. Rarely are they subject to pests or diseases.

Club Moss

RESURRECTION PLANT
Selaginella lepidophylla

A botanical curiosity comprised of a rosette of rather stiff, fernlike foliage. So sensitive to change in moisture is this plant that it has the remarkable ability to "fold up" its leaves when dry. Dehydrated, it forms a tight ball and can be "brought back to life" by placing in water.

Culture: Resurrection Plant

Temperature: Average household climate, not below 50°F (10°C).

Light: Partial shade to diffused sunlight.

Watering: Moist to dry. Appearance fluctuates with moisture content of soil.

Fertilizing: None needed.

Soil: Ordinary potting soil or grow in dish of water.

CLUB MOSS
Selaginella kraussiana

Easy rooting plant forms dense mass of yellow-green, spikelike foliage, which is excellent for a banking or carpeting effect in a terrarium. Attractive pot or basket plant provided humidity is adequate.

Culture: Club Moss

Temperature: Warm, 62-75°F (17-24°C), not below 50°F (10°C).

Light: Prefers shade, good in unbroken light of a north window all year. Humidity of 50% or more readily achieved by terrarium planting.

Watering: Constantly moist. Moisture condensation in terrarium makes for easier maintenance.

Fertilizing: Apply ¼ recommended strength every 4-6 months.

Soil: Humus-rich soil containing peatmoss or leafmold. Provide good drainage.

CUSHION MOSS
Selaginella kraussiana 'Brownii'

Cushiony mounds of densely clustered foliage are a vivid emerald green. This dwarf, tufted variety is a good foreground plant for taller terrarium plants.

Steps in Potting a Fern

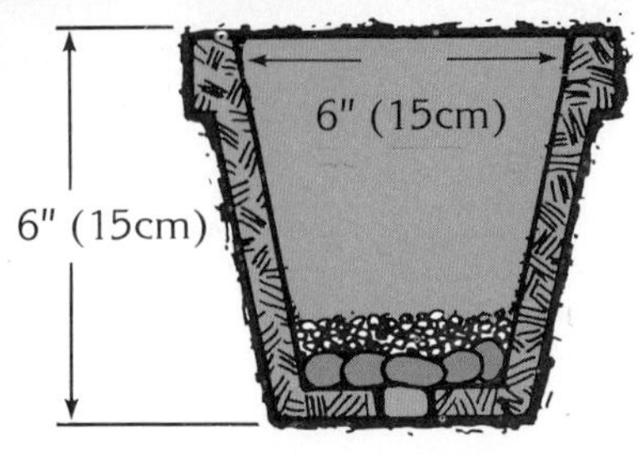

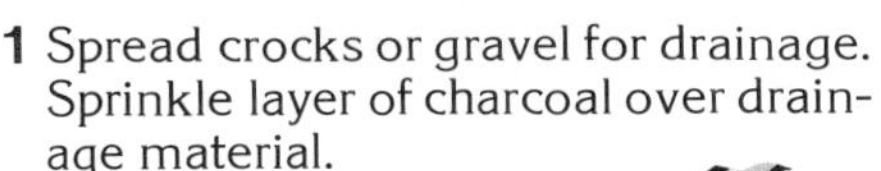

1 Spread crocks or gravel for drainage. Sprinkle layer of charcoal over drainage material.

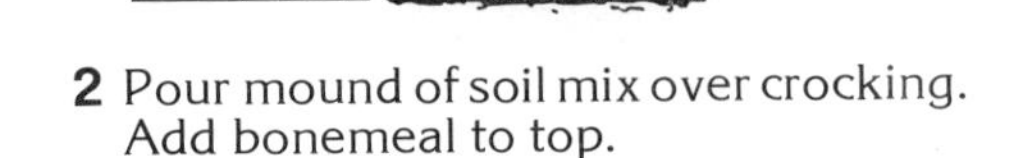

2 Pour mound of soil mix over crocking. Add bonemeal to top.

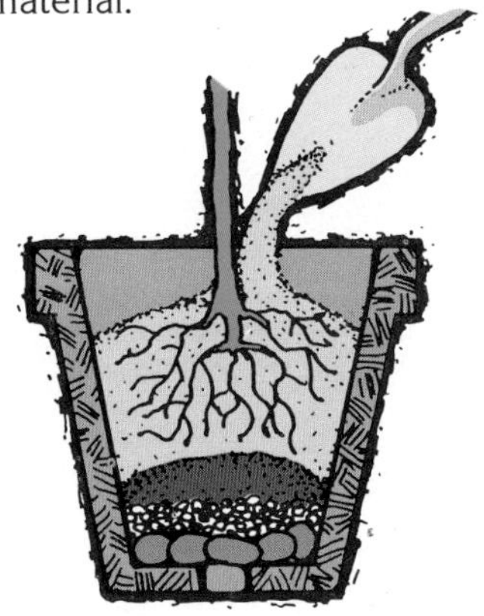

3 Center fern on mound of soil and fill in. If fern has an exposed rhizome or foot, do not bury this.

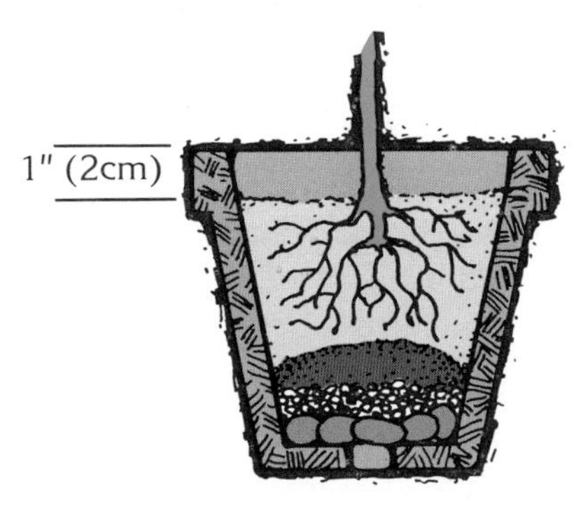

4 Secure fern by packing soil lightly and by rapping pot gently against potting bench. Leave 1" (2.5cm) open on top to receive water.

Containers and Drainage

Ferns do equally well in clay or plastic pots. Remember that plastic or ceramic (glazed) pots retain more moisture than clay, which is porous; therefore plants in plastic pots need less water. Tree fern fiber pots (osmunda) are good for epiphytes because they drain well, yet retain some moisture. They do not need to be lined. Wire or mesh baskets used for ferns should be lined with unmilled or natural sphagnum-moss strands, and then the soil mix filled in around the root ball.

Few plants survive if roots stand in water. Therefore it is important that broken clay pots or gravel be put in the bottom of all pots, those with, as well as those without, drainage holes. A layer of charcoal spread over the drainage material insures a sweet soil, and acts as a buffer against overwatering.

Because the majority of ferns do not have extensive root systems, they do well in fern pans or shorter pots, referred to in the trade as ½-pots and ¾-pots. Standard size pots, those whose height is equal to the diameter of the top, can also be used for ferns. However, it may be necessary to half-fill a standard pot with extra drainage material to decrease the depth when planting some kinds of ferns. (See notes for specific ferns.)

Do not overpot. Usually a pot diameter of 1 or 2 inches larger than the root ball is sufficient. This is important. Overpotting is unhealthy.

Propagation

Ferns reproduce not by seed but by special structures called spores. These are usually found on the underside of the fronds in tiny capsules (sporangia) that are grouped in clusters. They can be brown, black or even orange. Except for their neat, even arrangement, they look much like scale. (Do not confuse spores with scale—see Pests and Disease, page 76). If your fern does produce spores, pride yourself on having given it proper living conditions.

Can you obtain a fern from spores? Very likely, yes. In fact, your fern was probably grown from spores by a professional. Propagation of ferns from spores is a rewarding hobby that does require special techniques. Those techniques will not be discussed here, because there are simpler ways of acquiring new ferns for yourself or a friend.

Division

You cannot obtain a new fern by inserting a frond in water. It will not grow roots, but you can divide the roots. Division, cutting a section of root off with stems and fronds attached, is the simplest and quickest way to produce a new fern. Use a clean, sharp knife to divide the root clump. Don't be afraid that such drastic surgery will harm your fern. The parent will regenerate and produce more luxuriant growth in no time at all, and so will the offspring. After potting your newly divided plants, keep the soil moist and mist the plants. If wilting occurs, increase humidity by covering with a clear plastic bag held up and away from the fronds by thin stakes. Keep the new plants (divisions) away from strong light for a couple of days. Most ferns, in particular the Boston fern, can easily be divided.

Rhizomes

New fern plants can also be obtained by division of the creeping rhizome or foot-system of Rabbit's-Foot and Squirrel's-Foot ferns. Rhizomes are the visible, thick, fleshy stems that grow along the surface of the soil. With a sharp knife, cut through the rhizome. Make sure to include an end-foot or sprouting point in your cut. If the cut rhizome has roots, gently dig around and under it and remove the whole piece—roots, stems, fronds. Pot the root, leaving the severed rhizome exposed on the surface of the soil. Then treat as you would a divided plant.

Smaller rhizome pieces or rootless pieces without stem and fronds will also produce new growth in time. Place the rootless rhizome on top of moist soil (do not bury it) in a box, flat, or pot with good drainage. Keep it humid, shaded, and warm until rooting and/or evidence of sprouting (new fronds forming) is observed. A terrarium is a good place to start a new fern from a rhizome, root, or rootless section.

Buds

Some, like the Mother ferns, produce new plants called fernlets right on their fronds. Others, like the Hart's-Tongue fern, produce new plants if the stipe (frond stalk) is inserted as a cutting into a moist soil medium. These fernlets develop from knots in the tissue called buds. Fronds with buds can be anchored to the soil while still attached to the mother plant or frond; or they may be detached and planted separately. Keeping them attached to the mother plant assures the developing fernlets a good supply of water and food. If you choose to detach them, plant in moist, well-drained soil and provide ample humidity until the fernlets are large enough to be exposed to a drier atmosphere. About 2 years is required to grow an 8" (20.3cm) plant.

Terrariums

Ferns and mosses (selaginellas) are ideal plants for terrariums. Both like humidity and tolerate lower light. In fact, terrariums can be thought of as "humidity chambers" for maidenhair and other ferns that require higher humidity than is typical in the average home.

Choosing a Container The selection of a container for a terrarium is limited only by the imagination. Terrariums and kits can be purchased almost anywhere, but don't overlook the "free" terrarium in glass refrigerator jars or fruit-jar bottles. They make charming miniature terrariums for one or two plants and/or some moss. A 10-gallon aquarium can mimic a forest floor or woodland area with appropriate fern and moss plantings. Always strive to create a natural landscape in a terrarium, and consider how it will be viewed (from the top, from all sides?) before you plant. Avoid overcrowding. Each plant should be visible. Draw attention to the terrarium with a focal point — the tallest plant, a flowering plant, perhaps an African violet, or a stone, shell, or piece of driftwood. Any statuary should complement, not detract from the natural scene. Oversized statuary throws it off scale.

Preparing the Soil Layers Provide internal drainage so the soil will not become waterlogged. Build a drainage layer on the bottom, using ¼" to 1" (.64-2.5cm) of gravel. Spread a layer of charcoal over the gravel as a soil sweetener. A good soil mix to use is 1 part packaged potting soil, 1 part sand, and 1 part peat-moss. Perlite, which is white, may look unnatural in a terrarium. If you want to substitute it for sand, dye it brown with tea. You are now ready to select your plants.

Some Choice Terrarium Ferns Any Maidenhair fern, as *Adiantum 'Pacific Maid,'* or *Adiantum trapeziforme*; Mother ferns *(Asplenium daucifolium* and *bulbiferum)*; Button fern *(Pellaea rotundifolia)*; Boston ferns, smaller varieties, as *Nephrolepis 'Mini-Ruffles'*; Dwarf Leather fern *(Polystichum tsus-simense)*; Table ferns, as Silver Table fern *(Pteris ensiformis 'Victoriae')*; Squirrel's-foot ferns, as *Davallia fejeensis*; Moss *(Selaginella)*, as club moss or cushion moss.

A Few Points About Terrariums

- either glass or plastic is suitable.
- use only a clear, transparent container or one with a very light tint. Darker

tints reduce light and visibility of plants.
- keep soil moist, but do not overwater.
- supply adequate bottom drainage.
- apply water gently with a sprinkler or a baster. A stream of water will create pockets in the soil and disturb roots.
- allow moisture to evaporate from time to time, but some condensation is normal.
- do not fertilize, or plants will outgrow container too rapidly.
- give artificial or bright diffused sunlight, but avoid direct rays of sun.
- do not place the terrarium in a warm or hot place, as on top of a television set or radiator.

Enjoy your terrarium, but remember it is not meant to last forever. To extend its life, prune or replace overgrown plants. When simple pruning or replacement of one or two plants will no longer help the appearance of the terrarium, it is time to replant.

Pests and Disease

Ferns are relatively free from pests and disease. Too much water, poor drainage, and insufficient air circulation combined with too low temperatures and high humidity may cause gray mold (botrytis) on the soil and/or loss of fronds. Remove dead and injured fronds, and maintain a warmer, drier environment. Use a systemic fungicide (Benelate, Benomyl) as a drench on or around the infection. To prevent disease, avoid overwatering, water in the early part of the day, and space plants so as to insure adequate air circulation.

Scale is the fern's worst enemy. It is a small sucking insect encased in a hard, flat, brown shell. It is more likely to be found on the stalk than on the foliage. Don't confuse scale with the fern spores which appear in a regular pattern only on the undersides of leaves or along leaf margins. Adult scale is easily scratched off with a toothpick or cotton swab soaked in rubbing alcohol. Cut off and destroy badly infested fronds.

To destroy immature scale (those not yet housed in their protective brown shells), use 20% of the recommended strength of wettable Malathion powder. Apply with an atomizer or small brush. Later remove residue by spraying the leaves with water.

Mealy bugs, white cottony tufts or balls along the stalk or on fronds, sometimes attack ferns. Use the same method of control as with scale.

Red spider mites thriving in low humidity rarely attack ferns. Tiny white spots on fronds which later become mottled are an indication of these mites. To prevent, increase humidity. Kelthane sprays are safe to use on **some** ferns. Otherwise, remove adult insects with frequent sprays of water, and treat with a solution of wettable malathion powder.

Aphids, small, barely visible greenish or black insects, rarely attack house ferns. Remove them with a forceful spray of water and/or dip the plant in a dilute solution of malathion.

Fungus gnats, drain flies, and **springtails** look alike and resemble the common fruit fly. All three insects are commonly found in or around standing water in bowls or pebble trays, and in damp soil that has a high humus content. They are mostly a nuisance, but can spread fungi and bacteria. The best prevention is to keep plant areas clean. Periodically, empty water from pebble trays and disinfect them with chlorine bleach, (1-2 tbsp. per qt./1.5-3 ml per liter) To attack the larvae that may be in the soil, a soil drench can be prepared from wettable Malathion.

Ferns, especially Maidenhair, are sensitive to all chemical sprays containing petroleum distillates. The same is true for aerosol sprays, and neither should be used on ferns. Always read the manufacturer's directions and precautionary notes before applying any chemical to your ferns.

Cultural Problems of Ferns

Ferns have the same needs as other foliage plants grown indoors. If your fern starts to decline, check your growing conditions before you reach for an insecticide. Ill health is not necessarily the result of insect damage or disease. Check your soil and consider whether you have been giving too much or too little water. Most problems with ferns start there, or are the result of too dry air (insufficient humidity), too much or too little light, hot or cold drafts, insufficient fertilizer, or potbound roots. Specific problems are discussed under each fern.

Disorder	Probable Causes and Cures
Leaf margins turn brown	Air too dry — increase humidity Soil too dry — water more often and thoroughly Too much light — reduce light Salt damage — decrease use of fertilizers and leach the soil
Fronds turn yellow or brown	Check soil: if dry — water more often or more thoroughly; if moist — salt damage, leach the soil if wet — maybe poor drainage or simple overwatering Temperature too high; air too dry; insufficient fresh air, or minor nutritional (iron) deficiency
Brown spots on fronds	Overwatering or poor drainage. Light too bright.
Edges of fronds curling under	Insufficient humidity when fronds were developing
Fronds and leaflets falling off	Overwatering or poor drainage. Soil too dry Cold draft Insufficient light
Slow growth	Plant entering normal dormancy or rest period Soil too heavy Temperature too low Potbound root system Roots dying, caused by overwatering and/or salt damage

Disorder	Probable Causes and Cures
Fronds pale green or yellow	Overwatering Temperature too low Too little or too much light Fertilizers needed
New growth thin and pale	Too little light Too high humidity Too much nitrogen (N) in fertilizer
Few fronds or new fronds undersized	Plant needs repotting — check roots Plant needs fertilizer Air too dry Temperature too high Salt damage — leach soil Soil too heavy
Fern wilts	Check soil — if dry, water more often and more thoroughly If moist — salt damage, leach soil If wet, poor drainage or simple overwatering Plant needs repotting — check roots
Whole fern suddenly collapses	Roots damaged and are dead — caused by overwatering, fertilizer burn and/or high salt, cold or hot draft.
Rotting leaves and stems, gray mold on surface of soil	Due to fungi which attack where growing conditions are poor. Overwatering — especially in winter Letting water remain on fronds at night Poor ventilation, high humidity with too low temperatures

MAILORDER FERN SUPPLIERS

EXOTIC FERNS AND OTHER TROPICALS

Albert & Merkel Bros., Inc.
2210 S. Federal Hwy.
Boynton Beach, Florida 33435
(Send 50¢ for catalog)

Bolduc's Greenhill Nursery
2131 Vallejo St.
St. Helena, California 94574

Ferns & Foliage Gardens
1911 E. Alosta
Glendora, California 91740
(Send 13¢ for price list)

The Fern Farm
P.O. Box 364
Dallas, Texas 75221

Talmadge's Fern Gardens
354 G Street
Chula Vista, California 92010

HARDY FERNS AND WILD FLOWERS

W. Atlee Burpee Co.
Warminster, Pennsylvania 18974
(Free)

Ferndale Garden
705 Nursery Lane
Faribault, Minnesota 55021
(Free)

Van Bourgondien
245 Farmingdale Rd.
P.O. Box A
Babylon, New York 11702

"Fern Sources in the United States — 1972" by Donald G. Huttleston, printed in the **American Fern Journal**, 62 (1):9-15 available through The American Fern Society
Dr. Terry R. Webster
Biological Sciences Group
University of Connecticut
Storrs, Connecticut 06268
(Send 25¢ for reprint)

Index

Index continued

BACK COVER PHOTOGRAPHS:

Bottom left: **Hare's-Foot Fern,**
Polypodium aureum areolatum See p. 55

Center left: **Regina Wilhelmina Staghorn**
Platycerium bifurcatum 'Netherlands' See p. 53

Upper left: **Wild Athyrium,**
Athyrium species See p. 27

Bottom right: **Verona Lace Fern**
Nephrolepis exaltata 'Verona' See p. 41

Center right: **Five-Fingered Maidenhair,**
Adiantum pedatum See p. 21

Upper right: **Bird's-Nest Fern,**
Asplenium nidus See p. 25